PREPARING FOR CLIMATE CHANGE

PREPARING FOR CLIMATE CHANGE

Michael D. Mastrandrea
and Stephen H. Schneider

A Boston Review Book

THE MIT PRESS   Cambridge, Mass.   London, England

MIT Press books may be purchased at special quantity
discounts for business or sales promotional use. For
information, please e-mail special_sales@mitpress.mit.edu or
write to Special Sales Department, The MIT Press,
55 Hayward Street, Cambridge, MA 02142.

This book was set in Adobe Garamond by *Boston Review*
and was printed and bound in the United States of America.

Library of Congress Cataloging-in-Publication Data
Mastrandrea, Michael D.
Preparing for climate change / Michael D. Mastrandrea and
Stephen H. Schneider.
    p. cm.
"A Boston Review Book."
ISBN 978-0-262-01488-5 (hardcover : alk. paper)
1. Climatology. 2. Climatic changes. I. Schneider, Stephen H.
II. Title.
QC981.M424 2010
304.2'5—dc22

                                            2010023787

10   9   8   7   6   5   4   3   2   1

CONTENTS

Introduction

THERE IS GROWING WORLDWIDE MO-
mentum to address the problem of climate
change, one of the widest-reaching challenges modern society has faced. But we did
not reach our current level of global concern
without bumps and bruises along the way.

The natural greenhouse effect and its
intensification by human-induced (anthropogenic) emissions of greenhouse gases are
well understood and solidly grounded in
basic science. This conclusion is a robust
finding of the mainstream climate-science
community. Yet, despite the preponderance

of evidence, a number of interest groups—and some scientists—still do not accept the well-established evidence of the last 40 years of anthropogenic global warming.

Unfortunately, the media often treat these skeptics as credible experts, and they are given equal billing with mainstream scientists. One result is public confusion, which contributes to an already heated dispute. Climate change is not just an area of scientific study, but also a matter of public and political debate. Responding to climate change will fundamentally affect natural systems, energy production, transportation, industry, government policies, development strategies, population-growth planning, distributional equity, and individual freedoms and responsibilities around the world—in short, the well-being of human and ecological systems. Decisions on the scale and timing of climate policy will entail an array of costs and benefits for stakeholder communities with conflicting priorities. Moreover, all of this will play out in a background of varying degrees of knowledge, and thus inherent uncertainties.

Some of these uncertainties can be resolved by normal scientific investigations in the next decade or two. Others are almost guaranteed to remain until long after we are committed to cope with changes that can neither be predicted with high confidence, nor reversed after they are confidently detected. This poses a major challenge for planetary-scale governance of our development pathways.

Policymakers, lobbyists, financial interests, environmental advocates, and climate contrarians have struggled mightily to turn the weight of public opinion—and the funds controlled by it—in their preferred directions. Most mainstream scientists have countered with the methods at their disposal: research to increase understanding and predictive capacity, responsible reporting of research data, best-practice theory, international cooperation, and calls for policy consideration. Decision-makers, faced with myriad claimants of "truth," have come to rely on institutions that assess the relative credibility of the claims. Most countries use their own academies of science for assessments at national scales.

But the difficulties of international cooperation demand an international effort. For this reason, in 1988 the United Nations Environment Program and the World Meteorological Organization established the Intergovernmental Panel on Climate Change (IPCC). Every five to six years, the IPCC publishes its peer-reviewed, world governments–approved Assessment Report, which presents the best approximation of a global consensus on climate-change science.

Each report includes an assessment of the likelihood that its major conclusions will come to pass, and a rating of the authors' confidence in the science underlying that assessment. This practice clearly separates the more probable outcomes from those that are more speculative. Both experts and governments extensively review drafts of the reports during the development process, and a final Summary for Policy Makers (SPM) is approved in a "Plenary" process in which hundreds of government delegates work with the lead-scientist authors to determine precise wording. It is difficult to get all parties to agree on language, and the process inevitably eliminates outlier positions from both sides of the bell curve, but the consensus on the SPM allows "buy-in" from most national governments on the basic conclusions of the IPCC assessment reports.

Diverse Interests, Uncertain Outcomes

There is now overwhelming evidence for human-caused climate change. The science demonstrating a significant warming trend over the past century is settled. The most recent report—the Fourth IPCC Assessment Report (AR4), in 2007—called it "unequivocal." Moreover, it is essentially settled that the past four decades of warming largely have been caused by human activity—IPCC AR4 called it "very likely"—and that much more warming is in store for the 21st century given that emissions continue to rise. But how much warming can we expect, and how intense will the effects be?

On these questions, the scientific literature cannot provide the same level of confidence. The uncertainty estimates over how severe warming and its impacts will be by 2100 vary by a whopping factor of six. In part, this is due to uncertainty about the likely response of the climate system to the future trajectory of greenhouse-gas emissions. But a larger factor is uncertainty about the trajectory itself, which is dependent on future socioeconomic development and policy decisions that affect emissions.

The policy task, then, is to manage the uncertainty rather than wait an indefinite period to try to master it. This kind of risk-management framework often is employed in defense, health, business, and environmental decision-making. The IPCC, therefore, has focused on assessing scientific research detailing the threats posed by climate change at different magnitudes of future change, how likely those magnitudes of climate change are to materialize under various "business-as-usual" scenarios, and potential response strategies. These projections suggest that business-as-usual entails a variety of potential dangers.

The IPCC has been an important factor in motivating governments to consider reducing emissions of greenhouse gases. Not surprisingly, those whose financial interests rely on emissions have tried, usually unsuccessfully, to besmirch the credibility of IPCC science. Failing that, they have turned more recently to attacking IPCC processes and procedures, or individual scientists. These campaigns have been more successful.

For example, a small but highly politically damaging number of errors in IPCC conclusions were uncovered after the publication of the AR4 in 2007. Most notably, one conclusion was based on a weak, non-scientific reference that suggested a specific date—2035—for melting of Himalayan glaciers. There is currently no way to estimate with high confidence the levels of warming that would trigger this serious consequence or the rate at which it would unfold, even if set in motion. Given the uncertainties, no single number can be assigned any confidence—there must be a range of outcomes. But the erroneous conclusion remained in the Report undetected, and amid the fallout many missed the Report's correct conclusion that, according to high-confidence observations, the Himalayan glaciers were indeed melting.

Many in the media and nearly all of the opponents of IPCC conclusions attacked the credibility of climate science in general and of the IPCC in particular. Some even claimed that the errors were deliberate exaggerations designed to attract research funding. It is of course a legitimate news story that scientists make mistakes and that improved procedures to reduce error rates are needed. But few stories or attacks on the IPCC mentioned that this small number of errors appeared among thousands of pages of assessment and hundreds of conclusions that have not been challenged. In fact, the IPCC procedures include guidelines on the treatment of uncertainties intended in part to avoid such potential errors. In the vast majority of cases, the IPCC's guidelines worked as intended. The IPCC track record for accurately reporting the state of the science and the scientific confidence that can be attributed to various conclusions is unprecedented among assessment activities for complex systems. Certainly the worlds of finance, security, and health have nowhere near as high a percentage of unchallenged conclusions.

As already noted, significant uncertainties plague projections of climate change and its consequences. Science strives to overcome uncertainty through data collection, research, modeling, simulation, and other information-gathering approaches, and continuing research into the climate system will eventually reduce uncertainty about the effects of increasing atmospheric concentrations of greenhouse gases. But given the complexity of the global climate system, many decades' worth of high-quality data will have to be carefully analyzed.

Meanwhile, even the most optimistic business-as-usual emissions trajectory is projected to result in some potentially dangerous climate impacts for certain regions, sectors, and groups. That means we cannot avoid making policy decisions before significant uncertainties are resolved. Risk analysis—the scientific assessment of the consequences of potential outcomes and their probability of occurrence—is then distinguished from the more value-laden job of risk management—choosing how to hedge against the risks identified in the scientific-assessment process.

Extensive and sustained global action is required to cope with climate impacts already in the pipeline and to prevent even more damaging climate change in the coming decades. The aim is clear: reduce the growth of greenhouse-gas emissions and eventually bring those emissions significantly below current levels. In contemporary policy debates, efforts to achieve this goal are called *mitigation*.

It is also clear, however, that mitigation will not be enough to address the climate problem. Even with aggressive global efforts to reduce emissions, the earth's climate will continue to change significantly for many decades at least, due to past emissions and the inertia of social and physical systems. Significant impacts resulting from climate change are already evident, and they pose increasing risks for many vulnerable populations and regions.

Alongside mitigation, then, we also need policies focused on *adaptation*, on making sensible adjustments to the unavoidable changes that we now face. And we must coordinate adaptation with mitigation, as the success of each will depend on the other. Today's efforts to reduce emissions will, in due course, determine the severity of climate change, and thus the degree of adaptation required—or even possible—in the future. At the same time, a better understanding of the levels of climate change to which adaptation is difficult will help to shape our judgments about how much mitigation is required.

This book outlines the challenge society faces in addressing climate change in all its dimensions. We begin with an overview of the science of climate change and its potential impacts, continue with a discussion of strategies for responding to climate change—adaptation and mitigation—and conclude with a call for *bottom-up/top-down vulnerability assessment*, which brings together bottom-up knowledge of existing vulnerabilities and top-down climate-impact projections. Together these provide a transparent basis for informing decisions intended to reduce vulnerability, particularly adaptation decisions.

I

The Scientific Consensus

Since the second half of the nineteenth century, global temperatures have been on the rise. The increase in global average surface temperature, as estimated by the IPCC, is around 0.75°C (~1.4°F). Twelve of the thirteen years leading up to 2009 are the twelve warmest years on record. There is now overwhelming scientific evidence of a human fingerprint on this global warming.

Many impacts of warming can be—and have been—observed: the melting of mountain glaciers, the Greenland ice sheets and parts of the West Antarctic ice sheets, and northern polar sea ice; rising and increasingly acidic seas; increasing severity of droughts, heat waves, fires, and hurricanes (the intensity and/or frequency of extreme events can change substantially with small changes in average conditions); and changing lifecycles and ranges of plants and animals. The primary driver, particularly of the rapid warming since the 1970s, is emissions of greenhouse gases, such as carbon dioxide and methane, generated by human activities. The burning of fossil fuels is the greatest contributor of greenhouse gases, but agricultural practices, deforestation, and cement production also play a role.

The Warming Planet

The greenhouse effect and its intensification by human-induced emissions are well understood and solidly grounded in basic science. The potential of carbon dioxide in the atmosphere to trap radiant heat was proposed as early as 1827 by the French mathematician and physicist Joseph Fourier. In 1896 the Swedish chemist Svante Arrhenius dubbed this the *greenhouse effect*. Arrhenius was the first to argue that anthropogenic increases in the level of carbon dioxide in the atmosphere could significantly affect surface temperature.

So how does it work? Earth's atmosphere is moderately transparent to visible light. About half of the radiant energy from the sun penetrates the atmosphere and is absorbed by the Earth's surface. The other half either is absorbed by the atmosphere or reflected back to space by clouds, atmospheric gases, aerosols, and the Earth's surface. The absorbed energy warms the surface and atmosphere, which re-emit energy as infrared radiation. To stay in energy balance, the Earth must radiate back to space as much energy as it absorbs, but the atmosphere is much less transparent to infrared radiation.

Carbon dioxide and other greenhouse gases and clouds absorb 80-90 percent of the infrared radiation emitted at the surface and re-emit energy in all directions, both up to space and back toward the Earth's surface.

Thus, some infrared radiant energy is trapped, heating the lower layers of the atmosphere and warming the surface further. As it warms, the surface emits infrared radiation upward at a still greater rate, and so on, until the infrared radiation emitted to space is in balance with the absorbed radiant energy from sunlight and the other forms of energy coming and going from the surface (for example, rising plumes of convective energy, or evaporated water vapor that carries a great deal of latent chemical energy from the surface to the clouds where it is released in the condensation process).

The natural greenhouse effect makes our planet much more habitable—about 33°C warmer than it otherwise would be. But human activities are increasing the concentrations of greenhouse gases in the atmosphere directly and indirectly, thus intensifying the greenhouse effect. The indirect effect primarily stems from the extra evaporation of water from a warmed surface, a feedback that adds more water vapor—a greenhouse gas—to the atmosphere, warming the surface further. These amplifying influences are called *positive feedbacks in radiative forcing*, since the net effect of the addition of greenhouse gases when averaged over the globe is to trap extra heat, which in turn increases temperatures in order to restore energy balance. Greenhouse gases commonly emitted in human activities include carbon dioxide, methane, nitrous oxide, and a host of industrial gases such as chlorofluorocarbons that do not appear naturally in the atmosphere. Indirectly, humans also generate ozone in the lower atmosphere. The concentration of ozone, a health-damaging component of smog, is increasing with atmospheric warming and continued burning of fossil fuels.

These same activities—fuel combustion, and, to a lesser extent, agricultural and industrial processes—also produce emissions of aerosol particles. Many aerosols directly reflect incoming solar energy upward toward space, a *negative radiative forcing*, or cooling effect. Aerosol particles also affect the color, size, and number of cloud droplets, in aggregate, a negative forcing. Some dark aerosols, such as soot, absorb solar energy, a positive forcing if they darken the planet enough to cause more sunlight to be absorbed. Another indirect effect is soot falling on snow and ice, darkening it and thus accelerating melting. Many land-use activities, such as deforestation, contribute to greenhouse-gas emissions, a positive forcing, but they also can change the Earth's albedo, or reflectivity, in aggregate, again, a negative forcing. However, deforested surfaces may warm locally due to the removal of evapo-transpiring vegetation that cools the surface.*

The best available estimate of the combined influence of all human activities to date is strongly positive. Its magnitude is roughly equivalent to the positive radiative forcing of increased carbon dioxide concentrations alone, with the positive forcing of the non-carbon dioxide greenhouse gases and dark aerosols roughly offset by the negative forcing of direct and indirect aerosol effects and land-use changes, though the many uncertainties involved mean that precise estimates are not yet possible with high confidence. However, we can be highly confident that the overall effect is positive, and thus that human activities are contributing to observed warming.

---

* For more on radiative forcing, aerosols, albedo feedbacks, and other details, see the resources in the "Further Reading" section.

What besides human activities could be at work in the warming of the planet? Many natural processes affect the Earth's energy balance and therefore climate, which varied a great deal in the distant past. Aerosols ejected from large explosive volcanic eruptions can remain in the stratosphere for several years, all the while cooling the lower atmosphere by a few tenths of a degree. Changing solar output can alter temperatures by similar amount over the course of decades, and the sunspot cycle has a small, but discernible effect on solar output (~0.1 percent). Some scientists and interested parties champion these natural processes as the primary sources of warming in our own era. But natural processes alone do not cause a sufficiently sustained radiative forcing to explain more than a small fraction of the observed warming of the past 40 years. On the other hand, anthropogenic forces can explain a much higher fraction of what has been observed over that period.

Examining climates of the more distant past allows scientists to compare the current changes to earlier natural ones. Scientists use proxies that provide a window into those natural fluctuations. Proxies such as tree rings and pollen percentages in lake beds indicate that current temperatures are the warmest of the millennium and that the rate and magnitude of warming likely have been greater in the past 50 years than during the rest of this period. Ice cores bored in Greenland and Antarctica provide estimates of both temperature and atmospheric greenhouse gases going back hundreds of thousands of years, spanning several cycles of warmth (5,000-20,000 year "interglacials") separated by ice ages up to 100,000 years in duration. Not only do the samples indicate a strong correlation between temperature and atmospheric greenhouse-gas concentrations—particularly carbon dioxide and methane—the samples also indicate that current levels of carbon dioxide and other greenhouse gases in the atmosphere are far above any seen in at least the past 650,000 years. Ice cores also provide information about volcanic eruptions and variations in solar energy, furthering understanding of these natural forcing mechanisms described above.

There are many other lines of evidence of the human "fingerprint" on observed warming trends. To give one more example, the Earth's stratosphere has cooled while the surface has warmed, an indicator of increased concentrations of atmospheric greenhouse gases and stratospheric ozone-depleting substances rather than, for example, an increase in the energy output of the sun, which should warm all levels of the atmosphere. Combined, the present-day observations and the data provided by proxies have led the IPCC to conclude that it is *very likely* (there is at least a 90 percent chance) that human activities are responsible for most of the warming observed over the twentieth century, particularly that of the last 40 years.

Nevertheless, the future course of climate change is deeply uncertain because we don't know how much more greenhouse gases humans will emit or exactly how the natural climate system will respond to those emissions. Policy decisions can strongly influence the first source of uncertainty (future emissions), but will have little influence on the second (climate response to emissions).

Modeling Climate Change

This uncertainty means that projecting future climate change is a complex, imprecise task. There is a range of plausible futures. Using computer models that describe mathematically the physical, biological, and chemical processes that determine climate, scientists try to project the response of the climate to future scenarios of greenhouse-gas emissions. The ideal model would include all processes known to have climatological significance and would involve spatial and temporal detail sufficient to model phenomena occurring over small geographic regions and over short time periods.

Today's best models strive to approach this ideal but still rely on many approximations because of computational limits and incomplete understanding of climatically important small-scale phenomena, such as clouds. The resolution of current models is limited to a geographic grid-box of roughly 50-100 kilometers horizontally and one kilometer vertically. Because all physical, chemical, and biological properties are averaged over each grid-box, it is impossible to represent "sub-grid-scale" phenomena *explicitly* within a model. In other words, the specific climatic goings-on within the grid-box must be approximated.

But sub-grid-scale phenomena can be incorporated *implicitly* by a parametric representation. This "parameterization" connects sub-grid-scale processes to explicitly modeled grid-box averages via semi-empirical rules designed to capture the major interactions between these scales. Developing and testing parameterizations to assess the degree to which they can reliably incorporate sub-grid-scale processes is one of the most arduous and important tasks of climate modelers. The best models reproduce approximately, although not completely accurately, the detailed geographic patterns of temperature, precipitation, and other climatic variables seen on a regional scale, and can project changes in those patterns given scenarios for future greenhouse-gas emissions.

IPCC AR4, of which both of us were authors, includes climate-model projections based upon six "storylines," possible future worlds that come about under different assumptions about population growth, levels of economic development, and technological advancement and deployment. In one scenario, the IPCC assumes heavy reliance on fossil fuels and significantly increasing emissions during the century, and projects further global average surface warming of 2.4-6.4°C by the year 2100. In a second scenario, emissions grow more slowly, peak around 2050, and then fall, with expected warming of 1.1-2.9°C by the year 2100. The difference between the temperature ranges for the first and second scenarios reflects the influence of different trajectories for future greenhouse-gas emissions and climatic responses to those emissions: how much will temperatures increase for a given increase in concentrations (how sensitive is the climate to radiative forcing)? And how will the carbon cycle and the uptake of carbon dioxide by the ocean and by terrestrial ecosystems be altered by changing temperature and atmospheric greenhouse-gas concentrations?

These different projections for warming imply very different climate-change risks, affecting other climate variables (for example, precipitation patterns) as well as the likelihood of severe impacts. Warming at the high end of the range could have widespread catastrophic consequences and very few benefits, save the viability of shipping routes across an ice-free Arctic Ocean, or the possibility of expanded oil exploration in that sensitive region. Five to seven degrees Celsius of warming on a globally averaged basis is about the difference between an ice age and an interglacial period; in this case, the change would occur in merely a century or so rather than over millennia as in the paleo-climatic history of ice-age cycles not influenced by human activities.

Warming at the low end of the range (a few degrees Celsius) would be less damaging, but would still be significant for some communities, sectors, and natural ecosystems. Human civilization has grown in an age in which global temperatures were never more than a degree or two warmer than now, thus warming exceeding a degree or two is unprecedented in our entire historical experience. Indeed, some systems have already shown worrisome responses to the ~0.75 ° C warming over the past century. Alarmingly, actual emissions of the past ten years (except for a year or so of temporary decline during the economic recession of 2008-9) exceed the assumptions of even the highest of the IPCC scenarios, which were crafted in 2000. This suggests that large increases in greenhouse-gas concentrations are in store in the next several decades unless rapid action is taken to reduce emissions.

II

Impacts

MOST OF US THINK ABOUT CLIMATE in local terms. The Caribbean has great weather—warm days and cool nights, plenty of sunshine, blue skies. It's much nicer than dreary London or parched Dubai. All of these local conditions, however, are the products of an enormously complex global system in which myriad variables contribute to a diverse set of climates and ecosystems. That diversity has been relatively stable for the past several thousand years—until humans dramatically expanded their population size and economic activities. Now, major alterations to land surfaces, chemical composition of soils, air, and water and accelerating changes in global average temperature, even seemingly small changes, are upsetting that relative stability, affecting local conditions all over the planet.

The IPCC AR4 summarized many projected impacts of climate change for specific regions and highlighted "key vulnerabilities." These include the loss of glaciers, melting ice sheets, and other factors that produce rising seas, which could inundate low-lying coastal areas and small island nations around the world; escalating infectious disease transmission; increases in the severity of extreme events such as heat waves, storms, floods, and droughts; large drops in farming productivity, especially in hotter areas; the loss of cultural diversity as people are driven from their historical communities; and an escalating rate of species extinction.

Not Just Theoretical

Many of the types of problems discussed in the IPCC Report can be witnessed in their early stages today.

As glaciers melt, sea level rises and water in turn becomes scarcer in regions that depend heavily on glacier water during their dry seasons. In South America a significant fraction of the population west of the Andes could be at risk due to shrinking glaciers. According to a 2005 study from researchers at the University of San Diego, glacier-covered areas in Peru have shrunk by 25 percent in the past three decades. The authors note, "at current rates some of the glaciers may disappear in a few decades, if not sooner" and warn that fossil water lost through glacial melting will not be replaced in the foreseeable future. China, India, and other parts of Asia are also vulnerable. The ice mass in the region's mountainous area is the third largest on Earth following Arctic-Greenland

and Antarctica, and as its glaciers diminish in the coming decades, decreasing water supplies will affect vast populations. The Chinese Academy of Sciences has announced that the glaciers of the Tibetan plateau are vanishing so fast that they will shrink by half every decade. Researchers estimate that enough water permanently melts from them each year to fill the entire Yellow River.

While some worry about their dwindling water supplies, others, particularly vulnerable populations and those with little capacity to adapt, have begun to experience the direct health impacts of climate change acutely. For example, the increased frequency and intensity of heat waves put small children and the elderly at risk, especially where air conditioning is unavailable or unaffordable. Devastating events such as the 2003 European heat wave—now linked to the premature deaths of some 50,000 people—illustrate the dangers that exist even in developed countries. Increases in the frequency and/or intensity of floods, hurricanes, fires, and other extreme events are also troubling. The immediate effects of, say, wildfires are obvious, but the indirect impacts can be more damaging to health: smoke degrades air quality, exacerbating respiratory illnesses of millions in downwind areas.

In some regions—particularly the Arctic, where surface air temperatures have warmed at approximately twice the global rate—changing climate patterns are threatening entire ways of life. The island village of Shishmaref, off the coast of northern Alaska, has been inhabited for 4000 years. Its 600 current residents are facing the very real possibility of exile. Rising temperatures are melting sea ice, thereby allowing higher storm surges to reach the shore. Permafrost is thawing along the coast, increasing shoreline erosion and undermining homes and water systems. The absence of sea ice in the fall makes traveling to the mainland to hunt moose and caribou more difficult. Inuit hunters in Canada's Nunavut Territory report thinning sea ice, declining numbers of ringed seals, and new insect and bird species in their region. In the western Canadian Arctic, Inuvialuit are observing more thunderstorms and lightning—formerly very rare in this region. Norwegian Saami reindeer herders report that prevailing winds they rely on for navigation have shifted and become more variable, forcing them to change their traditional travel routes. Unpredictable weather, snow, and ice conditions make travel hazardous, endangering lives. The precise links of these local changes in weather patterns to climate change are difficult to establish, but the ill effects are illustrative of the broader risks of extreme events and changing climate patterns.

With regard to biodiversity, climate changes are having potentially irreversible effects on plant and animal habitats and lifecycles, forcing some species poleward or up mountain slopes, and hastening the arrival of certain biological events each spring. Depending on the severity of its impacts and the rates of response among different individual species, climate change could pull apart the natural functioning of existing plant and animal communities, making extinctions much more likely.

For example, over the past several decades, warming has led to the early arrival of some birds that migrate in the spring. If those arrivals are no longer in sync with the emergence of vegetation needed for nesting or hatching of bugs that are prey for these birds, then the interlocked life cycles of these co-dependent species can be disrupted.

Such disruptions are not only a threat to biodiversity, but also ecosystem "goods"—seafood, fodder, fuel wood, timber, pharmaceutical products, etc.—and "services"—air and water purification, flood control, pollination, waste detoxification and decomposition, climate moderation, soil-fertility regeneration, etc.

In addition to these well-understood effects of climate change, climate change could trigger "surprises." These are fast, non-linear climate responses, thought to occur when environmental thresholds are crossed. Some of these surprises could be anticipated. "Imaginable surprises" include the collapse of the North Atlantic thermohaline circulation (ocean currents)—which could cause significant and potentially rapid cooling in parts of the North Atlantic—and deglaciation of Greenland or the West Antarctic ice sheets, which would occur over many centuries (though would persist over many millennia), causing a considerable rise in sea level, threatening many coastal cities and low-lying coastal areas such as river deltas. But there is also the possibility of true surprises thanks to the enormous complexities of the climate system and the relationships, for example, between oceanic, atmospheric, and terrestrial systems.

III

Understanding Risk

ASSESSING CLIMATE SCIENCE, IMPACTS, and policy issues rarely involves certainties. Instead, we consider risks—potential outcomes associated with different levels of climate change, and the range of future climate change that could be induced by different levels of future emissions. In other words, what are the consequences, and what are the chances that they will be realized?

Assessing risk is primarily a scientific enterprise, but deciding which risks to tolerate and which to try to avoid—"risk-management"—is primarily a value-laden, normative activity appropriate to the political process. The climate problem is filled with deep uncertainties, uncertainties in both likelihoods and consequences that are unlikely to be resolved to a high degree of confidence before we have to make decisions about dealing with their long-term, and in some cases potentially irreversible, implications. These decisions often involve strong and conflicting interests and high stakes.

Philosophers Silvio Funtowicz and Jerome Ravetz have described such problems at the intersection of science and society that require decision-making under inherent uncertainty as "post-normal science." In Thomas Kuhn's "normal science," the practice is to reduce uncertainty through standard science: data collection, modeling, simulation, model-data comparisons, and so forth. The objective is to overcome uncertainty—to make known the unknown. New information, particularly reliable and comprehensive empirical data, may eventually narrow the range of uncertainty. According to this paradigm, further scientific research into the interacting processes of the climate system can reduce uncertainty about how the system will respond to increasing concentrations of greenhouse gases.

Post-normal science, on the other hand, acknowledges that while normal science continues its progress, some groups want or need to know the answers well before normal science has resolved the uncertainties surrounding the problem at hand. In that case, there will not be a clear consensus on all important scientific conclusions, let alone policies to reduce risks that will affect different stakeholders in different ways.

Scientists Educate—Society Acts

Decision-makers must weigh the importance of climate risks against other pressing social issues competing for limited resources. Some fear that actions to control potential risks might unnecessarily consume resources that could be used for better purposes, especially if impacts turned out to be minimal.

This can be restated in terms of *type I* and *type II* errors. If governments were to apply the precautionary principle and act now to mitigate risks of climate change, they would be committing a type I error if their worries about climate change proved exaggerated and anthropogenic greenhouse-gas emissions caused little dangerous change. If, on the other hand, policymakers chose to delay action until greater certainty could be established, and in the process stood by as serious damage occurred, they would be guilty of a type II error. Deciding which kind of error to avoid is not only a scientific activity (i.e., assessing risk), but also a value judgment (choosing which type of risk to face).

Such debate is critical to informed policymaking, and scientists regularly engage in it. For example, Working Groups 1 and 2 in IPCC AR4 had a type I/type II debate over sea level–rise projections. Working Group 1 scientists projected about one to two feet of rise over this century from thermal expansion of the oceans and one component of the melting of ice sheets in Greenland and Antarctica ("mass balance"). They chose to omit sea level rise contributions from another component of ice sheet melting—"dynamical melting" that is believed to be an important component of observed melting—because the existing set of ice-melt models were under-predicting the rate at which melting was actually being observed. Instead Working Group 1 added a caveat that the sea level–rise projections did not include the effects of dynamical melting. Since we in

Working Group 2 were required by governments to take a risk-management approach, we assigned a medium confidence ranking—one-third to two-thirds chance—to the conclusion that sea level could rise four to six meters over centuries to millennia.

In debates with Working Group 1 colleagues, we argued that paleoclimatic history and faster-than-predicted melting require us to estimate, but not with high confidence, the concerning potential of meters of rise in centuries, a relevant time frame for ports and coastal cities. Working Group 1 colleagues pointed out, correctly, that there is no scientific consensus on this conclusion. However, the important consensus in this case is not on one specific outcome, but rather on the *confidence* we have in the scientific basis for the range of possible outcomes. Scientists need to report even a 50-50 chance of meters of rise because the consequences would be severe.

Society, not scientists, should decide how to react to uncertain, but significant, risks. Therefore, we believe scientific information about the range of possible outcomes needs to be communicated to decision-makers, since what to do about the prospect of low-probability/high-consequence outcomes is a risk-management judgment that only society should make. The happy ending to this story is that the Working Groups agreed on a fair compromise that governments approved: risk of meters of sea-level rise in centuries to millennia. This type-I-versus-type-II-error debate ended up further informing governments about both the potential risks of climate change and how to frame arguments about it.

IV

*Preparing for
Climate Change*

EVEN THE MOST OPTIMISTIC BUSINESS-as-usual emissions pathway is projected to result in some dramatic, and potentially dangerous, climate impacts. Therefore, despite uncertainty over the future of climate change, we have to improve on the status quo. Faced with these grave risks, and great uncertainty, what should we do?

While we cannot know the precise temperature increase and impacts of a specific trajectory for future emissions, we do know a few things with confidence. We know that reducing emissions will reduce the level of

temperature increase that would otherwise occur, and thus reduce climate-change risks: that
is why mitigation is so important. But we also
know that further climate change will occur
no matter how quickly we are able to reduce
emissions. And we know that emissions are
increasing rapidly and are at higher levels than
assumed in the highest IPCC scenario. The
combination of historical and currently increasing emissions has locked in further warming for
many decades. In other words, climate change
is happening, and we need mitigation *and* adaptation. How can we get the right mix?

Mitigation and Adaptation

Mitigation and adaptation often are pre-
sented as trade-offs, as if pursuing one would
deflect attention and resources from the other.
But there is growing recognition that the two
policies must be complementary and concur-
rent. The Copenhagen Accord (produced at the

United Nations Climate Change Conference in 2009) sets 2° C above pre-industrial global surface temperatures as a threshold beyond which further warming is unacceptable. This target is informed by scientific research that examines the potential impacts of future climate change, but ultimately reflects a value judgment about acceptable levels of risk. This is about 1.25° C above current levels—a very challenging target, given that global emissions are still growing. This is a central reason why we see mitigation and adaptation primarily as complements: what cannot be prevented through mitigation must be adapted to; what we cannot cope with by adaptation, we must prevent.

Mitigation can keep warming on a lower trajectory by preventing some of the temperature increase that would otherwise occur if we continued with the high-emissions trajectory of business as usual. Some warming, however, will still be associated with a lower trajectory, and the impacts of this warming must be addressed by adaptation. Adaptation is a response to warming, not a means of slowing it. Delays in mitigation will lock in further warming, making it that much harder to adapt. Furthermore, due to the decades of inertia in both the climate and economic systems, the benefits of mitigation take time to materialize, so adaptation is essential in responding to near-term climate changes. Failure to adapt could be disastrous for many sectors, regions, and groups.

It is also crucial to understand that mitigation and adaptation yield fundamentally different benefits. Mitigation provides long-term, global benefits. A central challenge of mitigation policy, therefore, is to balance global factors. After all, responsibility for historical emissions, growth in current emissions, and capacity to reduce emissions vary widely among nations. The benefits of adaptation strategies are, in contrast, both more immediate and more region- and sector-specific. Given the wide range of climate-change impacts on different regions, groups, and sectors, the need for adaptation varies widely as well.

Avenues of Adaptation

The IPCC delineates two types of adaptation: autonomous and planned. Autonomous adaptation is not guided by policy; it is a reactive response prompted by the impacts of climate change. Consider a physiological example of autonomous adaptation: people who now live in warmer areas have acclimatized to those conditions and become less vulnerable to temperatures that would cause significant heat-related illnesses among people living in more temperate areas. Even so, there are limits to such adaptation, particularly if warmer temperatures spur increased use of air conditioning and therefore less acclimatization. Planned adaptation can be reactive too. For example, since the 2003 European heat wave, some countries have instituted more coordinated plans to deal with periods of extreme heat. Buying additional water rights to offset the impacts of a drying climate, or purchasing crop insurance where available, are reactive responses as well.

Any reactive adaptation almost certainly will not be fast or easy. Farmers, for example, may resist unfamiliar practices, have difficulties with new technologies, or face unexpected pest outbreaks. Moreover, the high degree of natural variability of weather may mask clear identification of emerging climatic trends. Suppose that in a certain area, slowly building climatic trends will generate much wetter conditions over time. But farmers faced with an anomalous sequence of dry years might easily mistake them for a new climatic regime and invest in maladaptive strategies such as increased water storage that becomes unnecessary, rather than flood prevention that will be critical in the long term. Adaptations to slowly evolving trends embedded in a noisy background are likely to be delayed by decades, as farmers and others attempt to sort out true climate change from random climatic fluctuations.

Another kind of planned adaptation—anticipatory or proactive—has greater policy potential. Anticipatory adaptation might include improving or expanding irrigation for agriculture, engineering crop varieties that are better able to cope with changing climate conditions, building sea walls to protect coastal infrastructure, and constructing reservoirs or implementing "greywater" recycling to improve water management by reclaiming wastewater from domestic activities.

One U.S. state that could benefit from anticipatory adaptation is California. With its Mediterranean climate of wet winters and dry summers, California relies heavily on melting snowpack stored in the Sierras for agricultural and urban water supply. Warming is expected to reduce the snowpack considerably—as more precipitation falls as rain instead of snow—and to melt the snow pack earlier in the year. Many of the actions mentioned above are being considered. The state might also take regulatory and political actions: connect protected lands to create migration corridors, set up networks to disseminate information about climate changes and potential adaptive actions, and create insurance mechanisms or support funds for disadvantaged and vulnerable groups that might not have the capacity to adapt on their own.

A generally effective near-term strategy likely will identify and pursue actions that not only address immediate threats, but also strengthen the ability to cope with natural climate variability. We are best served by anticipating more intense and/or more frequent extreme events than have been seen historically. Such an approach would build resilience as research continues to illuminate the severity and details of future climate change.

Avoiding severe impacts also will require long-term planning, such as investments in durable infrastructure in coastal zones or habitat protection for threatened or endangered species. In these cases, it is vital that we consider the full range of climate projections over the next century. And as the timeline lengthens, policy coordination becomes essential. Policymakers need to consider how adaptation policies will interact, both with each other and with attempts at mitigation. For example, certain adaptation options, such as recharging groundwater to increase water supply, may be energy-intensive, thus increasing emissions of greenhouse gases if that energy is generated from fossil fuels.

Anticipatory adaptation is an investment,

and most studies about its potential assume that countries and groups can afford it. Unfortunately, this is not universally true, especially for countries where development is a top priority. Several funds therefore have been established to help developing countries pursue adaptation measures, the best-known being the Marrakech Funds (established at the UN Climate Change Conference in Marrakech in 2001) and the Global Environmental Facility's (GEF) Climate Change Operational Programme, funded by world governments.

Yet, while these funds are promising, guidelines for determining which adaptation projects deserve funding are lacking. The GEF requires such projects to show "global environmental benefits," and the Marrakech Funds try to assure funding of adaptation to long-term climate change rather than to short-term climate variability. But it is difficult to assess adaptation projects on these grounds because they are local (and therefore bring local, rather than global, benefits) and will likely improve an area's ability to adapt both to climate change and climate variability. Moreover, there is not enough funding. The Copenhagen Accord tried to correct this. It includes a commitment from developed countries to provide "adequate, predictable and sustainable financial resources, technology and capacity-building" to support the implementation of adaptation actions in developing countries, with plans to raise nearly $30 billion over the next three years and $100 billion per year by 2020. This is a step in the right direction, if it is implemented.

Planned and autonomous adaptation both have their limits, which is why we have been stressing the need for adaptation and mitigation in tandem. Sensitivity to changing climate conditions may be higher than currently estimated. Without significant mitigation of greenhouse-gas emissions, warming and the intensity of impacts are likely to exceed the coping capacity of adaptation measures in many sectors and regions.

A third, more drastic form of response to climate change is geoengineering. Schemes to modify environmental systems themselves or control climate have been promoted for more than 50 years in order to increase temperatures in high latitudes, increase precipitation, decrease sea ice, create irrigation opportunities, or offset potential climate change, among other objectives.

As a bulwark against climate change, various proposals recommend injecting iron into the oceans to promote algae growth, introducing sea-salt aerosol in the marine boundary layer, or spreading dust in the stratosphere or positioning mirrors in space to reflect solar energy and offset heat trapped by increased greenhouse gases. In a similar vein, a variety of managed relocation strategies have been proposed in order to prevent extinction of species that are unable to adapt independently to climate change.

These approaches—geoengineering, and what might be called ecoengineering—attempt to offset the effects of one global-scale manipulation of the Earth system (climate change) with another large-scale manipulation of physical or biological systems. Unsurprisingly, such manipulation may have unintended consequences. For example it is hard to justify saving a species by relocating it to new areas, thereby making it an invader in other habitats.

Geoengineering advocates claim their methods are cheaper and easier to implement than mitigation strategies slowed by foot-dragging governments and the lack of international agreements on long-term emissions reductions. But skeptics question whether any geoengineering scheme would work as planned without side effects, and whether the long-term international political stability and cooperation needed to maintain such schemes is attainable. Moreover, geoengineering risks transnational conflicts; such activities may produce—or be perceived to produce—damaging climatic events, and thus provoke political conflict.

Geoengineering represents a desperate attempt to deal with climatic impacts: understandable, but not what the situation demands as a first response. In reducing risks, nothing can substitute for the hard work of aggressive mitigation combined with anticipatory adaptation.

Equitable Solutions

Even with an optimal mix of mitigation and adaptation, the results may still be unfair. The most vulnerable groups are often the most marginalized and therefore the least able to influence decisions. Hence, policies often cater to powerful special interests—the coal industry; the United States, China, and other wealthy countries—at the expense of the more needy. To avoid overlooking already-marginalized groups when forming local, national, and international climate policy, decision-makers need to consider the effects of actions (and inactions) on the distribution of people's well-being and the sustainability of other species.

In a framework of distributive justice, disadvantaged countries and groups should be prioritized. Inequitable impacts can occur both from the direct effects of climate change and from the differential impacts of climate policies on the poor. Thus, good governance in the realm of climate policy requires both protecting the planetary commons by managing emissions and vulnerability, and dealing in fairness with those most disadvantaged by either climate impacts or by the effects of climate policies. The drivers of the problem—generally richer countries—can make payments to those who have contributed less—generally poorer countries—and these payments can be fashioned for fairness and political cooperation.

After the weak Accord at Copenhagen, many have suggested that the UN consensus process is too unwieldy to produce greenhouse-gas reduction targets muscular enough to avoid dangerous climate change. These parties seek to work around the UN process with privately negotiated deals among the main players such as the United States, India, China, the European Union, Japan, Russia, Mexico, and Brazil. At Copenhagen President Obama negotiated just such a deal with China, India, Brazil, and South Africa. Many nations subsequently signed on, but with reluctance and even annoyance, as they were not parties to the intense eleventh-hour bargaining.

These side deals could be effective in cutting the carbon output of the industrial emitters. At the Davos World Economic Forum in

January 2010, there were repeated calls to abandon the UN process in favor of deals among coalitions of the willing—and the capable. But if these "mini-lateral" negotiations become the norm, who will support adaptation and sustainable development in the poorest countries? That is why the UN process must remain the primary vehicle for collection and transfer of resources to nations that cannot meaningfully access mini-lateral mitigation deals. Structuring and managing a dual system of UN negotiations and mini-lateralism, side by side, will be a challenge for world leaders in the years to come.

V

*A New Way to Assess
Vulnerability*

IN OUR FINAL CHAPTER, WE HIGHLIGHT *vulnerability assessment* as an important tool to inform the development of climate change policies, particularly adaptation strategies.

Vulnerability often is defined in terms of three components: exposure, sensitivity, and adaptive capacity. *Exposure* refers to the degree to which a system experiences stress and the nature of those stresses: the frequency and intensity of heat waves in a given location, the level of the sea. *Sensitivity* refers to the degree to which a system is affected or modified by that exposure, and varies across different regions, populations, and sectors: the elderly and those without air conditioning are more susceptible to ill effects of heat waves; flat coastlines are more sensitive to rising seas than are steep ones. *Adaptive capacity* refers to the ability of a system to adjust to change, in terms of expanding the range of impacts with which it can cope, reducing its sensitivity to the changes, or both.

Mitigation reduces vulnerability by reducing exposure, while adaptation reduces vulnerability by turning adaptive potential into adaptive capacity, thus reducing sensitivity. The distinction between adaptive potential and adaptive capacity is critical. We know now that the vulnerability of New Orleans to a direct hit by a Category III hurricane was much higher than was widely believed prior to Katrina (though a small subset of academics and engineers had warned of this outcome for decades and were ignored). Adaptive potential was quite high—for example, levees could have been strengthened in advance—but this potential was not realized, and therefore adaptive capacity was low. In general, adaptive capacity is related to the level of development in a country. But events such as Katrina, which primarily affected poor citizens, and the 2003 heat wave in Europe, which primarily affected the elderly, highlight the vulnerability of specific populations and regions, even within highly developed nations.

Linking Assessment and Decision-Making

Assessing vulnerability to climate change is a complex task. It requires analysis of historical and current exposure and susceptibility to climatic conditions and their related impacts, projections of future impacts in the context of alternative socioeconomic development paths, and an evaluation of how well different adaptation strategies will do at reducing vulnerabilities. Detailed understanding of the affected sectors, communities, and management systems; the interactions of non-climatic stressors with a changing climate; and each system's ability to respond to changing conditions are often lacking. There is a critical need for research that couples climate projections with studies of vulnerability that focus on specific economic sectors (agriculture, services, manufacturing, etc.), regions, and groups. And these need to be generated in close communication with relevant stakeholders.

Decision-makers want understandable information about climate change risks. In particular, planners and managers in various sectors seek climate information that can support adaptation-related decision-making, provide straightforward estimates of uncertainty, and serves the needs of decision-makers in specific sectors. Such knowledge is ideally co-produced through sustained stakeholder-scientist interactions.

This interaction is crucial because, on its own, more information about projected climate changes and impacts does little to alter on-the-ground decision-making processes. A study investigating climate-change awareness and preparedness among coastal managers in California reported that most managers do not use weather, climate, or sea level–rise data in current decision-making, and that managers want more information on climate risks but only in a form that fits "seamlessly" into existing procedures.

Recommendations for adaptation actions based on scientific research often fail this test. For example, a 2009 study of 22 years of scientific literature on biodiversity conservation found hundreds of calls for adaptation of conservation practices to address climate change, but few recommendations with sufficient specificity to inform actual operations.

Decision-makers need concrete strategies and case studies that illustrate how and where to link research agendas, conservation pro-grams, and institutional practices. If the goal is to turn scientific analysis into policy action, then stakeholders and scientists must connect at all stages of the process: problem-detection, design of adaptation and mitigation plans, and implementation.

A Better Way: Bottom-up/Top-down Vulner-ability Assessment

To date, most climate-impact assessments have been top-down. They emerge from global-climate models, with all their attendant un-certainties. Model projections are based on the range of greenhouse-gas emissions asso-ciated with alternative future scenarios, but these often fail to account for the socioeco-nomic trends associated with each scenario, and the potential impact of these trends on vulnerability. The link between vulnerability and development is recognized by the IPCC and elsewhere, but beyond this recognition the literature is sparse.

Success in adaptation to climate change will come from the mating of top-down and bottom-up assessment. Scientific projections are most useful when joined with the intimate knowledge of existing vulnerabilities that stakeholders possess. Assessments based on biophysical (top-down) versus social (bottom-up) vulnerability provide complementary information, and comprehensive assessment of vulnerability to rapid climate change is impossible unless they are integrated. Detailed bottom-up studies provide understanding of the structural, institutional, psychological, financial, legal, and cultural frameworks of affected sectors, communities, and management systems. They can teach us much about our ability to cope with both a changing climate and non-climatic stressors that might worsen its effects.

Thus, the challenge is to develop integrative methods and to employ the resulting knowledge in order to inform decision-making. Again, this challenge can be met only with direct partnerships between stakeholders and scientists—social scientists who perform vulnerability assessments and climate scientists who speak clearly about what they do and do not know and how this information can be useful on the implementation end.

We call this approach—linking scientists with stakeholder experts in specific regions, sectors, or populations—*bottom-up/top-down vulnerability assessment*. It requires at a minimum the following three steps.

First, assess historical and current exposure and sensitivity to a wide range of climatic conditions and resulting impacts, both experienced (e.g., property damage or loss of life) and perceived (e.g., heightened sense of danger, loss of public trust). Second, assess existing adaptive capacity, decision-making processes (e.g., how fast can policies or behaviors change? how extensively?), and communications infrastructure. These steps together help reveal thresholds of exposure that would prove challenging for a particular system to adapt to, and thus provide a basis for defining current and future vulnerability thresholds associated with climatic exposure.

Third, integrate these bottom-up local assessments of vulnerability with top-down projections about climate change and socioeconomic development to examine the likelihood of exceeding such vulnerability thresholds identified in the bottom-up analysis as a function of top-down scenarios. Projections that reflect uncertainty in future climate change (like high-, medium-, or low-emissions pathways) can be employed to calculate the likelihood of crossing these thresholds of exposure. Development pathways (that exhibit different levels of societal adaptive capacity and vulnerability) can be employed to examine how exposure and sensitivity may change over time, and thus how vulnerability thresholds based on climatic exposure may change in the future.

Bottom-up/top-down vulnerability assessment provides a more transparent basis for tackling the challenges of climate change. It also enables managers to tailor adaptation strategies closely to the vulnerabilities of specific communities—say, indigenous peoples in the arctic or farming women in developing countries—and natural systems.

Our approach is not perfect. Because it relies on comparison of potential outcomes to past and present experience, it may not reveal important thresholds of vulnerability that have no contemporary or historical analog. The climate system is not so clear-cut; conditions may change faster than expected and novel combinations of stressors can produce surprises.

But the limits imposed by surprises are faced comparably by all approaches.

In addition to the uses already described, our method can be useful in establishing the roles of adaptation and mitigation. Consider the following scenario. Regional experts all over the world evaluate the well-being and vulnerability of local systems in an attempt to discern limits to adaptation. Scientists collect their findings and discover that the thresholds they establish cluster around a particular level of temperature increase. These data, collected in service of adaptation programs, also inform the mitigation debate about avoiding "dangerous" climate changes. In this sense, we argue that adaptation assessment becomes a complement to mitigation planning, not simply a trade-off as it is so often framed.

The Present Global Challenge

Given the uncertainties in climate science and impact estimates, we believe we must reduce considerably the rate at which we add to atmospheric greenhouse gas levels. This will give us more time to understand climate risks and to help develop lower-cost mitigation options, while making climate surprises less likely. Greenhouse gas–abatement policies will provide incentives to invent cleaner, cheaper technologies, and developed countries should aggressively lead that effort, both because of their historical contribution to the problem and because of their greater capacity to help.

Simultaneously, the needs of developing countries and marginalized groups should be accommodated through coordinated adaptation and mitigation actions. Developed countries should shoulder this burden as well, as required by the United Nations Framework Convention on Climate Change. When developing countries say they will not join mitigation efforts until they catch up with developed countries in per capita emissions, and some developed countries assert that they will not abandon fossil-based energy generation that props up their economic growth, we face real, potentially catastrophic environmental danger.

We will need international negotiations and bargaining to help the developing world leapfrog the traditional technologies of growing economies—like massive coal burning or dramatic increases in individual car use. With cooperation and political will, lower-emitting technologies such as electric vehicles can be built, and alternative-energy sources tapped, at much faster rates.

Slowing down pressure on the climate system and addressing the needs of marginalized countries and groups are the main "insurance policies" we have against potentially dangerous, irreversible climate events and the injustices that inevitably will accompany them. As the world struggles to fashion fair and effec-

tive forms of mitigation, adaptation, too, will
be essential if we are to avoid the worst consequences of climate change.

FURTHER READING

Science and Policy

• *Climate Change 2007: Working Group I Contribution to the Fourth Assessment Report of the IPCC.* The Intergovernmental Panel on Climate Change. Cambridge: Cambridge University Press, 2007.

The most thorough, authoritative source available. Volume 1, "The Physical Science Basis," covers the scientific understanding of climate change, past, present, and future. Volume 2, "Impacts, Adaptation, and Vulnerability," discusses regional and sector-specific impacts of climate change, as well as adaptation and vulnerability. The final volume, "Mitigation of Climate Change," presents general and sector-specific mitigation options and costs, as well as mitigation scenarios.

• *Climate Change Science and Policy.* Stephen H. Schneider, Armin Rosencranz, Michael D. Mastrandrea, and Kristin Kuntz-Duriseti, Eds. Washington, D.C.: Island Press, 2009.

Detailed and comprehensive—but accessible—reference on the science, impacts, and politics of climate change, with options for economic and energy policy.

• *Geo-Engineering Climate Change: Environmental Necessity or Pandora's Box?* Brian Launder and J. Michael T. Thompson, Eds. Cambridge: Cambridge University Press, 2010.

Overview of the potential benefits and risks of geo-engineering.

• *Global Climate Change Impacts in the United States.* Thomas R. Carl, Jerry M. Melillo, Thomas C. Peterson, and Susan J. Hassol, Eds. New York: Cambridge University Press, 2009.

Comprehensive, readable survey addressing what climate change could mean for the United States.

• *What We Know About Climate Change.* Kerry Emanuel. Cambridge, Mass.: The MIT Press (a Boston Review Book), 2007.

Overview of the basic science of climate change and how the current consensus has developed.

Public Affairs and History

• *The Discovery of Global Warming.* Spencer R. Weart. Cambridge, Mass.: Harvard University Press, 2003.

History of climate science and the study of climate change.

• *Fairness in Adaptation to Climate Change.* W. Neil Adger, Jouni Paavola, Saleemul Hug, and M. J. Mace, Eds. Cambridge, Mass.: The MIT Press, 2006.

Unique and detailed collection focusing on justice and adaptation to climate change.

• *Field Notes from a Catastrophe: Man, Nature, and Climate Change*. Elizabeth Kolbert. New York: Bloomsbury USA, 2006.
Primer on the consequences of climate change, told through descriptions of impacts observed around the world.

• *Science as a Contact Sport: Inside the Battle to Save Earth's Climate*. Stephen H. Schneider. Washington, D.C.: National Geographic, 2009.
Frontline account of the scientific and public debates on understanding and dealing with climate change.

Online Resources

• Climatechange.net
Overview of climate science, impacts, policy, and debates in the public arena.

• Realclimate.org
Essays and commentaries on "climate science from climate scientists," intended for journalists and the interested public.

ACKNOWLEDGMENTS

ANY BOOK IS THE CULMINATION OF THE work of many individuals, and this one is no exception. The authors gratefully thank the editorial team at *Boston Review*, particularly Deborah Chasman and Simon Waxman, who provided invaluable support and unhesitating editorial streamlining of our sometimes-wordy prose, and who were critical contributors to the production of this book and our *Boston Review* article with which this book originated. We also thank our editors at the MIT Press, Laura Callen and Clay Morgan, for all their publication efforts.

This book also draws from a spectrum of our research and other activities, and many thanks go to our colleagues and friends who have provided comments and advice, specifically on earlier versions of this book, and who have influenced our broader perspectives captured here. Notable among them are Patricia Mastrandrea, Terry Root, and Nicole Heller. Finally, we wish to thank our families for their irreplaceable support in all we do. They provide us with deep and ongoing happiness: Annabelle Louie, David Mastrandrea, and Patricia Mastrandrea, and Terry Root, Adam Schneider, Becca Cherba and grandson, Nikolai Cherba.

ABOUT THE AUTHORS

MICHAEL D. MASTRANDREA is Deputy Director, Science at the Intergovernmental Panel on Climate Change (IPCC) Working Group II, and Assistant Consulting Professor at the Stanford University Woods Institute for the Environment. His work has been published in *Science Magazine* and *Proceedings of the National Academy of Sciences*, and he is co-author of chapters on key vulnerabilities and climate risks and on long-term mitigation strategies for the 2007 IPCC Fourth Assessment Report. He also serves on the Editorial Board for the journal *Climatic Change* and is co-editor of *Climate Change Science and Policy*.

STEPHEN H. SCHNEIDER was Melvin and Joan Lane Professor for Interdisciplinary Environmental Studies and Professor of Biology at Stanford University. From 1973 to 1996 he was a scientist at the National Center for Atmospheric Research. A member of the National Academy of Sciences, he consulted for federal agencies and seven presidential administrations.

Schneider was Coordinating Lead Author and part of the Synthesis Report writing team for the 2007 IPCC Fourth Assessment Report. That year, he and his IPCC colleagues were awarded a joint Nobel Prize. He was founder and editor of *Climatic Change* and authored or edited hundreds of scientific papers, books, and other writings, including *Science as a Contact Sport: Inside the Battle to Save the Earth's Climate* and *Climate Change Science and Policy*. He passed away suddenly in July 2010.

BOSTON REVIEW BOOKS

Boston Review Books is an imprint of *Boston Review*, a bimonthly magazine of ideas. The book series, like the magazine, is animated by hope, committed to equality, and convinced that the imagination eludes political categories. Visit bostonreview.net for more information.